The Great Guilt
that causes the Deaf Effect

Jeremy Griffith

**Watch the video of this presentation at
www.HumanCondition.com/GG**

OR

Scan code to view

The Great Guilt that causes the Deaf Effect by Jeremy Griffith, with Addendum by Refentse Molosiwa

Published in 2022, by WTM Publishing and Communications Pty Ltd (ACN 103 136 778) (www.wtmpublishing.com).

All enquiries to:
WORLD TRANSFORMATION MOVEMENT®
Email: info@worldtransformation.com
Website: www.humancondition.com or www.worldtransformation.com

The World Transformation Movement (WTM) is a global not-for-profit movement represented by WTM charities and centres around the world.

ISBN 978-1-74129-073-8
CIP – Biology, Philosophy, Psychology, Health

Filming and editing by James Press & Tess Watson.
Cover image: *Fishermen at Sea* by J.M.W. Turner (1796)

The drawings by Jeremy Griffith, copyright © Fedmex Pty Ltd (ACN 096 099 286) 1991-2023.

Contents

FIX THE WORLD™
HUMANCONDITION.COM

With the real problem of the
human condition finally solved we
can now ACTUALLY fix the world!

Created by J. Griffith & G. Salter © 2020 Fedmex Pty Ltd

Background

Jeremy Griffith is an Australian biologist who has dedicated his life to bringing biological understanding to the dilemma of the human condition—the underlying issue in all human life of our species' extraordinary capacity for what has been called 'good' and 'evil'. Jeremy has published over ten books on the human condition, including:

— *Beyond The Human Condition* (1991), his widely acclaimed second book;

— *A Species In Denial* (2003), an Australasian bestseller;

— *FREEDOM: The End Of The Human Condition* (2016), his definitive treatise;

— *THE Interview* (2020), the transcript of acclaimed British actor and broadcaster Craig Conway's world-saving interview with Jeremy about his book *FREEDOM*;

— *Death by Dogma* (2021), which presents the biological reason why the Left and its Critical Theory threatens to destroy the human race; and

— *The Shock Of Change* (2022), about managing the change from ignorance to enlightenment of our human condition.

This booklet, ***The Great Guilt that causes the Deaf Effect***, is a transcript of a presentation Jeremy gave in 2022, which can be viewed at www.humancondition.com/GG. It serves as an important companion piece to *THE Interview*.

Jeremy's work has attracted the support of such eminent scientists as the former President of the Canadian Psychiatric Association Professor Harry Prosen, the esteemed American ecologist Stuart Hurlbert, Australia's Templeton Prize-winning biologist Professor Charles Birch, the Former President of the Primate Society of Great Britain Dr David Chivers, Nobel Prize-winning physicist Stephen Hawking, as well as other distinguished thinkers such as the preeminent philosopher Sir Laurens van der Post.

Jeremy is the founder and patron of the World Transformation Movement (WTM)—see www.humancondition.com.

The Great Guilt
that causes the Deaf Effect

[1] **Jeremy Griffith**: Welcome everyone to our Sydney World Transformation Movement Centre. It's Sunday 26 June 2022.

[2] We have some very special guests here today—Lucas Machlein from our WTM Centre in Nice, France, who's come to spend some months with us, learning the trade of how to promote this precious understanding of the human condition; also Fabiana Hargreaves de Costa, who's presently living in Norway with her five-year-old daughter Maia, but is from Brazil where she has started a WTM Centre in Rio de Janeiro.

[3] We also have some of the stars from our WTM Melbourne Centre: Ari & Desi Akritidis, their son Alex and daughter Nicoletta. Alex is also going to stay on for a little while to learn the trade.

[4] One precious person we're missing from our Sydney team is our Patron Tim Macartney-Snape who's currently on one of his

mountaineering treks to the remote peaks of Pakistan, but Tim is represented here by his partner, Stacy Rodger. My precious brother Simon Griffith is also out of town at the moment, and another key precious founding member, Tim Watson, is watching this meeting online. I also want to mention that Professor Harry Prosen is with us in spirit because the first anniversary of his death at the age of 90 was last Tuesday. I miss him heaps.

[5]Of course, the other very, very, very special people are Annie Williams beside me here and all my brothers and sisters here in the Sydney WTM Centre who have all been through so much to bring this project to where it is today on the threshold of lift-off!

- - - - - - - - - - - - - - - - -

[6]I want to use this opportunity to bring as much understanding as I possibly can to the problem our project suffers most from, which is the initial difficulty people have taking in or 'hearing' discussion of the human condition—which is what we refer to as the 'Deaf Effect'. The better everyone can understand the Deaf Effect, the faster our project will move forward and the faster the world will be saved from the frightening looming threat of terminal psychosis and our species' extinction—and as I have explained in my books, that threat is very real.

[7]So, this talk is about the Deaf Effect and how it's stalling this world-saving project, so let's bring some really deep understanding to that most critical of all issues.

- - - - - - - - - - - - - - - - -

[8]After watching *THE Interview*, then this presentation, and then *The Great Transformation*, on our home page, the next talk we advise people to watch is titled **'Your block to the most wonderful of all gifts'**.

[9]The obvious first thing to be asked about the title of that next talk is what is this **'most wonderful of all gifts'**? Well, once you

understand this information you will find that it actually finally brings relieving understanding to every aspect of human life. As WTM founding member Tony Gowing, who is sitting beside Annie here, says (and this comment appears in my book *FREEDOM: The End Of The Human Condition*): '**If you look at any of the problems in the world with any degree of honesty…they are each in a depressingly dark state, but with this understanding of the human condition that darkness is completely turned around into the most glorious, happy and light-filled situation imaginable…In the world before this explanation there were no answers, there was no meaning, no direction, no real understanding—I had no real idea what to do in the world, no framework of reference, no idea about the meaning of existence at all…But now I have complete understanding of the world.**' (paragraph 1256 of *FREEDOM*)

[10] So, that's an example of this wonderful gift. And there have been many, many comments like Tony's of the effects of being able to understand the human condition, but just to quote another one—this is from another founding member, Sam Belfield, that he posted on our Facebook Group this week about my new Addendum 2 in my booklet *Death by Dogma*. Sam wrote: '**without these insights I would just be lurching from denial to being utterly confused, angry, depressed and distressed about what is happening to Western civilisation, with nil awareness of the deeper implications for humanity…This handful of paragraphs is so clarifying and enlightening and relieving.**'

[11] Yes, the ability to explain the human condition just unlocks the most amazing ability to look into every aspect of our lives at last.

[12] So '**the most wonderful of all gifts**' is the understanding of the human condition that finally brings relieving insight into every aspect of human life. As to how it is able to do this, what I am first going to explain about the Deaf Effect will, I think, make it very clear why we couldn't explain anything truthfully and thus effectively until now. This is because being able to understand the human condition is what makes *all* the difference when it comes to understanding ourselves and the world.

[13] So the question screaming out to be answered is, 'What could possibly be **'our block'** to this **'most wonderful of all gifts'** of finally being able to understand everything?' Well, as I briefly mentioned, the problem is that most people struggle to take in or 'hear' discussion of the human condition. And this is obviously a very serious **'block'**, because if your mind can't take in or hear what's being said you're clearly not going to be able to discover **'the most wonderful gift'** of the healing understanding of every aspect of human life that the explanation of the human condition makes possible.

[14] To illustrate this, an American woman [May Gibbs] (who is shortly starting a WTM Centre) experienced this Deaf Effect when she said, **'I tried for months to get through the reading material but my thought at the time was that it was super tedious and boring.'** Saying it was **'super tedious and boring'** is a typical Deaf Effect response. As I'll hopefully make very clear, the subject of the human condition has historically been so unbearably confronting and depressing that most people's minds just don't want to engage with it, and so to cope their minds defensively and dismissively, in effect, say, 'I'm not interested in what you're talking about, it's just super tedious and boring, meaningless rubbish as far as I'm concerned!'

[15] Significantly, <u>when this woman persevered listening to and reading the information and eventually got through the Deaf Effect</u> she was able to discover this **'most wonderful of all gifts'** of being able to understand the world—so much so that she said, **'I find it laughable now…[how deaf I was because] here I am today…obsessing over the Freedom Essays, videos and Facebook Group posts…It has brought such peace to my life and I have a burning desire to get it to whoever will listen.'**

[16] So the Deaf Effect is incredibly real and is blocking access to this most wonderful gift of being able to understand ourselves.

[17] Fabiana, I know you struggled with the Deaf Effect; I remember reading you saying something to the effect that if it wasn't for wanting to help your daughter you wouldn't have <u>persevered</u>, persisted long enough to get through the Deaf Effect. Is that what your experience was when you struggled with the Deaf Effect?

[18] **Fabiana Hargreaves de Costa**: Yes, because when I came across this understanding I was having a lot of personal problems in my life and I desperately wanted something to guide me specifically on how to solve my problems. I couldn't see the macro level of all of this. And especially being a mother is a struggle, you feel that crazy love and you can't provide it to your child properly. You're just mad and crippled with the world. But once you understand this, it's just magical.

[19] **Jeremy**: Yes, but when you first tried to read *FREEDOM* you found it really difficult?

[20] **Fabiana**: Yes, because I wanted to solve the specific problems that I was feeling at that moment, so I put *FREEDOM* down because it wasn't helping because I couldn't access what it was saying. For one year I wasn't reading *FREEDOM* and then suddenly when I connected, I couldn't get enough of it. I was reading and listening to you 10 hours a day and it was just like a cry of relief and happiness and joy.

[21] **Jeremy**: So you had to 'swim' for a year, and then you finally did get back to it and could get through the Deaf Effect and discover how useful it is?

[22] **Fabiana**: Yes.

[23] **Jeremy**: And Ari, I know your brother Sam was telling you for many years about this information and I read somewhere where you said, '**When I eventually read Jeremy's book** *A Species In Denial* **not one word of it made sense to me.**' As you said yesterday, '**As soon as I read that I'm competitive, aggressive and selfish my brain turned off because I'm not any of those things—let alone a 2 million years corrupted human!**'—but now you have got through the Deaf Effect you and your whole family can't get enough of these explanations. Is that right? Is that a summary of your Deaf Effect experience?

[24] **Ari Akritidis**: Yeah, yeah. I'm still blown away at that Deaf Effect, Jeremy, that a number of years later I'm reading the same words and sentences and paragraphs and pages that are as beautiful as anything can be, and that I read that same thing a decade ago and I got zero, zilch; not a single word could penetrate my denial, you know, and that blows me away. But I was extremely deaf and just in such denial. And I said to my wife a few days ago, I would have read those words that '**I'm competitive, aggressive and selfish**' and said, 'That's not me, I'm none of those things'; I've been so focused on

being a good bloke and a good dad and a good brother and a good everything and completely in denial that there's anything wrong with me or the world and, yeah it's extraordinary, Jeremy.

[25] **Jeremy**: Yes, this Deaf Effect is very real, which is what I'm going to go on and try to explain and make very clear as to why.

- - - - - - - - - - - - - - - - -

[26] So the next question is, what exactly is it about the subject of the human condition that causes this initial Deaf Effect? The 10th video at the top of our home page that new people are recommended to watch is titled *'What exactly is the human condition?'* So that should be relevant in explaining what this problem is: what *is* the human condition? I think in hindsight what I said in that 10th video is a bit too indirect. I started by saying the human condition is 'the riddle of why we humans are competitive and selfish not cooperative and loving', and 'it's the issue of "good and evil" in our make-up', and even, 'it's the issue of "why we are the way we are"', but I think I should have got straight to the point and said that the human condition is the incredible guilt and shame we humans experienced when we became conscious and started being competitive, selfish and aggressive when our instinctive self or soul expected us to behave cooperatively, selflessly and lovingly—in other words, in the opposite way. The human condition is the guilt we feel for having corrupted our soul. That's what the human condition *really* is. This shame, this historic shame, which is what my work is all about explaining—why we corrupted our soul.

[27] I want to now explain as clearly as I possibly can this immense guilt and shame that arose when we became conscious and able to understand cause and effect and as a result started to wrest management of our lives from our dictatorial instincts, which, as I explain in my introductory interview with Craig Conway and in all my books, is what caused us to become psychologically upset and competitive, selfish and aggressive.

[28] So to start at the beginning of this horrendous saga for us humans, I want everyone to imagine what it was like when we first became conscious some 2 million years ago. How confusing and bewildering the world must have been when we consciously 'woke up', as it were; we became aware, and we looked around and started trying to understand everything. What was the meaning of thunder and lightning, did it represent some sort of attack on us? Why was the sun taken away at the end of each day and we were then given a pale substitute for it (the moon) surrounded by little eyes looking at us (the stars)? Why aren't animals kind to each other like we are? Why are crocodiles so damn nasty?, etc, etc.

Sensitive reconstruction of *Homo habilis* by acclaimed palaeoartist Élisabeth Daynès, suggesting the emergence of consciousness

[29] So these are the sort of questions we asked when we 'woke up', but because we've never been able to explain why we corrupted our soul, we haven't been able to stop to think about or try to imagine what it was like when we first became conscious some 2 million years ago and started thinking and watching and trying to make sense of everything around us. (As explained in pars 705-707 of *FREEDOM*, 2 million years ago is when our large association-cortexed, thinking brain first appeared in the fossil record of our ancestors.)

[30] The world was certainly a big mystery. But what I want to point out about that situation—and this is really interesting and very important—is that <u>we could cope with the confusion because we had our loving soul to look after us.</u>

[31] Bewilderment at this early stage of being conscious wasn't a problem because we had so much room in ourselves, so much kindness and generosity and love for each other and for the world around us. We could cope with mystery and hardship, and so our bewilderment about the meaning of everything wasn't too distressing. Children, if they grow up in a natural, nurtured, loving environment, are full of excitement and wonder, not full of fear and distress—I mean they're little consciousness-dawning beings—and so when the human species was still innocent, even though conscious thought was underway, we were like little children throughout our whole adult lives. We were doing lots of thinking but we were happy living together and living with nature around us. That is what life was like during the early hundreds of thousands of years of the development of conscious thought. We were still full of love and enthusiasm and happiness. Again, not being able to admit that we were once innocent because we couldn't defend why we corrupted that state, denied us the ability to think about what it was like to actually become conscious, and so forth. But we can now see that we were still in the arms of our soul that looked after us, if you like, and this all-sensitive and all-loving past was just the most wonderful existence.

[32] Now—and this is all-important—this wonderful existence didn't last. Eventually, as the upsetting search for knowledge developed further, that situation completely changed. Imagine the absolute horror when, for some reason that we had absolutely no understanding of, we started to become competitive, aggressive and selfish—behaving in a way that was completely at odds with our instinctive self or soul that only knew to behave cooperatively, selflessly and lovingly! [Again, see *THE Interview* for the explanation of why we became psychologically upset and competitive, selfish and aggressive]. Since we humans today have learnt to live in denial of our corrupted condition, it's difficult for us to connect with the guilt and shame and horror that we are actually living with, but the shame and guilt and frustration about why we became seemingly awful beings has been truly astronomical—as I'm now going to try to make very clear.

[33] So becoming conscious led to a lot of bewilderment about our world—the lightning and the sun disappearing and so forth—but none of those bewilderments were anything like as troubling as the issue of our corrupted condition. In fact, that issue was *so* troubling the agony of it was beyond anything we could bear.

[34] Sure we could come up with the excuse that we were just being like other animals, always fighting and aggressive—this is the false 'savage instinct' excuse I've talked about [see Video/Freedom Essay 14]—but initially that excuse didn't work because our instinctive self or soul, the 'voice' of which is our conscience, was letting us know we should be being cooperative and loving; we hadn't yet blocked out its voice, repressed our soul and its awarenesses into our subconscious, as we do now. We knew something had gone terribly, terribly wrong—that we were starting to behave appallingly, and we had absolutely no idea why. So that's another thing we've got to now start trying to immerse ourselves in—what it was like to *start* to become upset humans, behaving in total violation of our innocent soul.

[35] So, yes, we were starting to feel dreadful about ourselves, full of shame and guilt—completely at odds with the wonderful all-sensitive and all-loving world of our instinctive self or soul; in fact, we were immensely lonely beings who had in effect been booted out of a paradisical Eden, condemned as horrible monsters on Earth—what other conclusion could we come to other than we were dreadful beings?!

Adam and Eve cast out of Paradise, from *Old Testament Stories*, pub. Society for Promoting Christian Knowledge, London, c.1880

[36] Now, the following is one of Australian cartoonist Michael Leunig's wonderful cartoons that beautifully captures some of the horror of our situation where we were, in effect, thrown out of our soul's innocent Garden of Eden world—and how upset that made us. (see par. 274 of *FREEDOM*)

Cartoon by Michael Leunig that appeared in Melbourne's *The Age* newspaper on 31 Dec. 1988

Drawings by Jeremy Griffith, with deeply appreciative deference to Michael Leunig, Jul. 2009

[37] So we've got Adam and Eve eating the fruit from the Tree of Knowledge, which is a metaphor for becoming conscious, and then you see the guardian angel of the Garden of Eden throwing us out! And you can see that their expressions are actually quite

revealing. Eve is quite appalled, while Adam says, 'Bugger you'. So he's starting to get defensive and she's really distressed and he's starting to think about getting even, getting more and more angry. Finally, Adam gets the chainsaw out and just tears the whole place down, which is sort of what humanity has done. And *then* he sets fire to the whole joint and burns it to smithereens, which is again kind of what we've done. So this is a little metaphor of what happened—how humans were thrown out of innocence and felt ashamed and became defensive, angry and retaliatory.

[38] There's an extra little row I've drawn to add to the bottom of the cartoon, because now that we have found the redeeming and reconciling understanding of ourselves we are able to call the guardian angel back and show him that insight, which is presented in *FREEDOM*, that explains why we were good and not bad, to which the guardian angel starts crying in sympathy and takes us back to the Garden of Eden—so we're on our way home again. But you can see that Leunig's cartoon is a lovely metaphorical story of our shame and how it's affected us, how it upset us.

[39] Yes, we have been immensely, immensely lonely—feeling that we were just garbage on Earth and that everybody and everything hated us. It's hard for us to see our situation because we are now so practiced at denial, but in order to properly understand the Deaf Effect we need to try to imagine just how lonely our situation has actually been.

[40] The Biblical prophet Isaiah described our situation truthfully when he said: **'justice is far from us, and righteousness does not reach us. We look for light, but all is darkness; for brightness, but we walk in deep shadows. Like the blind we grope along the wall, feeling our way like men without eyes…** Truth [i.e. understanding of our corrupted condition] **is nowhere to be found'** (Isa. 59). Yes, as the prophet of our time, and now Nobel Laureate for Literature, Bob Dylan, sang, **'How does it feel to be on your own, with no direction home, like a complete unknown'** (*Like a Rolling Stone*, 1965).

[41] So what I'm trying to do is immerse everybody in a truthful rendition of the human journey of how horrible it was when we

started to corrupt the innocent, loving existence we once had, when
we had no idea why on Earth that was happening, and that the
shame was astronomical and the loneliness and wretchedness of our
existence terrible, because it felt like we had violated everything
fundamental about our world.

[42] William Turner's painting *Fishermen at Sea* captures something
then of the astronomical heroism of the human race for struggling
for *2 million years*, which is the time our species has been fully
conscious, through this terrible, terrible lonely darkness of guilt-
stricken bewilderment and seeming evil badness and the feeling
that left us with that we are no-good, utterly meaningless creatures.

J.M.W. Turner's *Fishermen at Sea*, 1796

[43] It's a very powerful picture. It's got this huge storm surging in
the darkness and there's people hunkered down in the middle of a

small boat trying to look after each other against this overwhelming world of condemnation. It's a very, very powerful image. You can see the people huddled together in the boat—we had to be our own friends because we were no longer a friend of our soul and the rest of the world associated with it. We have been very, very, *very* alone beings.

[44] Now, the following archaeological discovery is really interesting and evidences something of just how lonely our species has been. This is a photo of a settlement that was created some 12,000 years ago at a place called Göbekli Tepe in Anatolia, Turkey. It is the oldest permanent human settlement anywhere in the world, so it is absolutely amazing to have found and to think about. It possibly or even probably arose with the advent of agriculture that enabled people to transition from a hunter-gatherer existence to a sedentary life.

© Vincent Musi/Nat Geo Image Collection

[45] Now I want you to look at these T-shaped, sometimes 10-foot/3-metre high pillars that populate every enclosure. They clearly represent people, the arms on the side of the pillars continue down to hands that wrap around above the waistband with an animal skin serving as a loincloth.

© Vincent Musi/Nat Geo Image Collection

[46] Anthropologists aren't sure how to interpret these pillars, but I think it's clear they represent ancestors, and that what these Neolithic people were doing was surrounding themselves with the comforting presence of their ancestors in order to counter the utter loneliness of being condemned as evil beings.

[47] And this agony of our corrupted, lonely condition would have rapidly developed following the advent of agriculture and sedentary living, because moving from a hunter-gatherer existence, where we were surrounded by innocent nature all day long, to a village environment where people were now living in close proximity, greatly compounded the increase in upset in everyone. I've written about this in my books, where I quote the historian Manning Clark pointing out that **'The bush** [wilderness] **is our source of innocence; the town is where the devil prowls around.'** (see pars 941-942 of *FREEDOM*)

[48] To me, this next photo of the Bushmen of the Kalahari sitting together (from Lee and DeVore's *Kalahari Hunter-Gatherers*, 1976) is very like

the picture of the ancestor pillars that form the big circles found at Göbekli Tepe. The Bushmen are sort of sitting in a circle with their backs to nature because innocent nature was hating our upset state and so we huddled together and just looked after each other instead. As I explain with this photo of the Bushman in *FREEDOM*, innocent nature was so condemning of our upset we 'got even' with it by hunting down and killing innocent animals. Hunting was all about killing animals for their innocence's implied criticism of us, *not* about getting food, as we've been taught, which women's gathering actually supplied. In fact, this photo is titled *Telling The Hunt*; telling how 'I smashed some innocent animals to death, ha, ha, take that!'

© Laurence Marshall

[49] So yes, this picture of the Bushmen all looking inwards is very like the picture of the T-piece sites. If we look again at that first T-piece picture, you see they're all forming a circle, and you can see another circle in the background. All these ancestors are looking inwards.

[50] So I'm saying that surrounding ourselves with our ancestors was a reaction to the incredible loneliness of our lives when we couldn't understand why we had become seemingly evil monsters—again, like the fishermen heroically huddled together in their boat riding out the terrible storms around them and within them.

[51] Anthropologists don't recognise that our species was originally innocent, so they are not able to make sense of anything really. They can't begin to connect with the psychological predicament of our lives, so they're 'failure trapped' in their attempts to make any real sense of anything. Our species' original state of innocence is *such* a fundamental truth that to try to make sense of our world while denying it was like trying to understand how a car works while being determined not to look under the bonnet!

[52] As Professor Harry Prosen said about my interview with Craig Conway, it is **'the most important interview of all time because it turns all the conventional knowledge about human existence on its head with its recognition of the original cooperative and loving innocence of our species'**. Yes, *everything* starts to make sense once we can admit our species' original state of innocence, which we now can because we've found the good reason why we corrupted our soul—but nothing really makes sense when you can't admit that truth; it's just one big world of lies and bullshit, which is the atrociously dishonest world we have been living in. It *is* astronomically dishonest. As everyone in this room has discovered, solving the human condition, explaining why we are good and not bad, opens up a whole new world of truth—which, again, all begins by admitting our species once lived in an innocent loving state.

[53] No one has ever talked the way I am today, and in my books, freely about our species' original innocence and all the insights that gives us into the human journey. This is the first true description of us, which is why it is all so astonishingly interesting.

- - - - - - - - - - - - - - - - - -

[54] Since this is one of those interesting insights, I might also mention the occurrence of so-called 'Goddess' or 'Venus' figurines that have also been found in these Neolithic ruins in Turkey, like this one that was found at Çatalhöyük.

[55] As I've explained in chapter 5 of *FREEDOM*, nurturing is what created our moral soul, which means nurturing was the priority throughout our species' early development, and it was only after we became conscious and the human condition emerged that the priority shifted from being matriarchal to being patriarchal, but matriarchy didn't give in to patriarchy for a long time.

[56] As I explain in paragraph 810 of *FREEDOM*, the *extremely* regal stature of the very well-nourished figure seated on her throne of

cheetahs show just how powerful and in control of their societies women were right up to recent times, because these early settlements in Turkey around 10,000 years ago were clearly still matriarchal, still led by women.

[57] This situation where women seemingly held power in the Palaeolithic and Neolithic eras was a case of 'delayed ownership'—a situation where the new owner, patriarchy, wasn't able to take over because the old owner, matriarchy, refused to relinquish power. As I explain in *FREEDOM*, women are soul sympathetic, not ego sympathetic, so sooner or later 'ego sympathy', support for our species' upsetting battle to find knowledge, ultimately for understanding of ourselves, which was men's responsibility, had to take over. The explanation in chapter 8:11B of *FREEDOM* of the relationship between men and women is very important for everyone to read and understand.

[58] So I think it's very interesting that patriarchy hadn't taken over back then, which is not that long ago. And, by the way, I think we can observe in this well-nourished figure that we hadn't learnt that there was any problem with upset-driven over-eating back then. It took us a while to learn what's wrong with the idea that if food is enjoyable, and there is plenty of it, then why not enjoy it! The world of upset, distressed humans was a whole new existence!

- - - - - - - - - - - - - - - - -

[59] I might also mention that anthropologists don't know why the people at Göbekli Tepe avoided putting faces on the head pieces of their pillars—but I think with our ability to admit the truth of our lost state of innocence we can work out why. To engrave the faces of our beloved ancestors would have required depicting something of their tortured human condition when we just wanted their

comforting presence, not their agony. I wrote about how our soul couldn't and didn't want to draw our alienated faces in *FREEDOM* in paragraph 834.

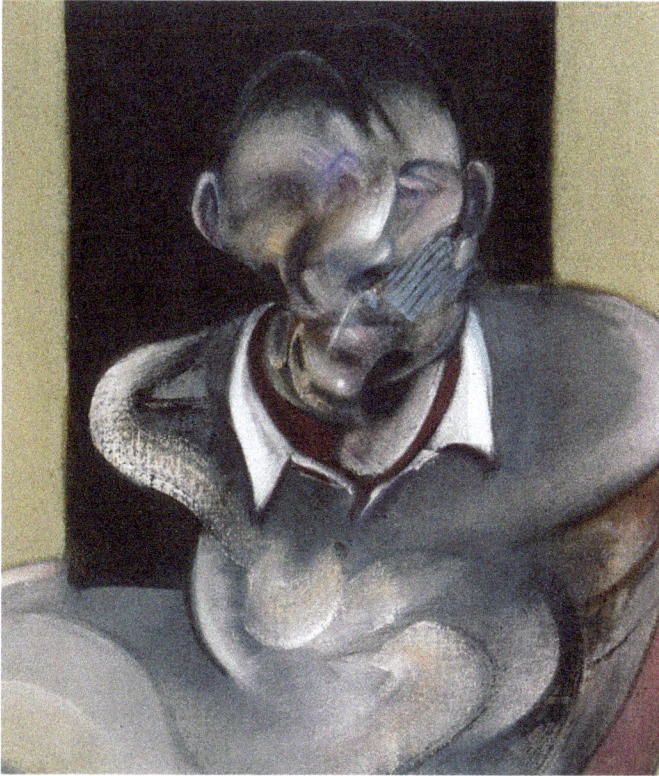

Detail from Francis Bacon's *Study for self-portrait*, 1976

[60] This painting by Francis Bacon from paragraph 124 of *FREEDOM* gives a true representation of how alienated we humans really are, so it makes sense that these Neolithic people didn't want to start reminding themselves of this, of how tortured we really are!

[61] Interesting to me is how incredibly empathetic my soul is when I just let it draw, especially draw happy soulful things, like I did for these people embracing each other, which was such an instant scribble when I did it, it shocked me how completely empathetic it was. Susan [WTM founding member Susan Armstrong], I think you were there when I sat down and did that quick scribble.

[62] **Susan Armstrong**: I was. It was incredible, it took about two seconds.

[63] **Jeremy**: Yeah, I scribble and then I stand back, because I can't draw for nuts normally, but if I just tune right into my soul it's magic what it can produce. [Note, later in this talk Jeremy explains his nurtured and sheltered innocence that has allowed him to access his/our instinctive self or soul.]

[64] Consider the soulful empathy in the following drawing I did of a mother and infant. I mean, how good's that! That is *so* tender. And you can see with her arms, I just threw them in, I didn't even try very hard, but her expression is so loving.

[65] Or the soulful empathy of this drawing I did of Christ [below]; I drew it because I know what he was like, which was an innocent, natural, alienation-free person, **'the Lamb of God'** as it says in the Bible (John 1:29), and that's what I tried to capture. I think it's a remarkably empathetic drawing.

[66] I know my soul can draw upset humans as well, like I did for this drawing of the oppressive effects of egocentric fathers on their children. (see par. 1011 of *FREEDOM*)

The child is turned into either a 'I-won't-back-down', 'must-win', power addict, or a psychologically crippled, broken person as an adult.

[67] But I think it's clear my soul likes drawing soulful humans, not upset humans, and I think that's what was going on with these T-pillars. The Neolithic people didn't want to show the upset in their ancestors, so they just left the top of the pillars blank, without faces. But I think the big blocks used for the heads do beautifully signify that these ancestors were a very big, comforting presence in their life—the pillars are very stylistically effective—they just didn't want to include reference to their ancestors' psychosis, so they left their faces blank. But the thinking amongst some anthropologists is that even though they can't explain the absence of faces, the figures were able to be identified by the carving of the person's totem or

symbol; in this case, by the fox on this pillar—as occurs amongst Australian Aboriginals where each person is given a totem of an animal that they're not allowed to hunt or eat.

[68] I should mention there have been a few carvings of faces found in these ancient settlements, but anthropologists think they are depictions of masks, and many ancient cultures do use masks in exorcism rituals for valving off upset.

Carvings of faces from Göbekli Tepe area

[69] So that all illustrates how lonely we humans have felt, feeling condemned and ostracised by the whole world—and thus so in need of comfort from those ancestors who bravely came before.

- - - - - - - - - - - - - - - - -

[70] To get back on track with what we are talking about. When humans became conscious we were certainly bewildered by the world—lightning and thunder, who stole the sun at the end of each day, why are animals so savage, etc, but the biggest bewilderment by a million miles was what occurred when we became conscious— why, when we'd all been so cooperative, loving and selfless, had we become competitive, aggressive and selfish? Worse, at times *ferociously* mean and vicious towards each other, even sadistic. For some completely unknown reason we had become brutal monsters from hell who had turned all that was good and wholesome into a dystopia of horror—we'd become the most vile, despicable creatures on Earth; so awful we felt we shouldn't exist. I mean animals are pretty ruthless, but they're not mean or sadistic like humans can be—which all means that thinking about how we corrupted such a magic life was utterly and completely depressing. A life of utter happiness had turned into a life haunted by depression.

[71] Sure we make huge efforts to put on brave faces and try to think positively, and to look positive, but the truth is that depression has been the real and dominant characteristic of human life since the latter part of our 2-million-year search for knowledge.

[72] You see, every second of every day, as Tony Gowing is always telling us, people are trying to maintain a positive outlook, they're trying to stay positive all the time, every moment. And the way they posture, everything's designed to hold at bay this deep psychological insecurity that I'm finally explaining. But that defensive, pretentious, 'I'm fine', 'Everything's good', as Ari described his

outlook, is all to counter what's really going on underneath, which is this deep insecurity that we can now understand is the horror of having become corrupted when we were originally innocent.

- - - - - - - - - - - - - - - - -

[73] So, at this point it might be good to hear some honest descriptions of how insecure we really are, how fearful we *really* are of the issue of the human condition.

[74] This is a description of the unbearable depression that the philosopher René Descartes felt when he tried to confront the horror of his corrupted condition: **'So serious are the doubts into which I have been thrown…that I can neither put them out of my mind nor see any way of resolving them. It feels as if I have fallen unexpectedly into a deep whirlpool which tumbles me around so that I can neither stand on the bottom nor swim up to the top.'** (par. 624 of *FREEDOM*)

[75] And this is another person's description of what he experienced when he tried to confront the human condition: **'I felt the worst fear I have ever known. Fear doesn't even go close to expressing it. What do you suppose you do when you find the most fearful thing you'll ever encounter is yourself.'** (par. 1185 of *FREEDOM*)

[76] Yes, as the psychoanalyst Carl Jung said, **'When it [our shadow] appears…it is quite within the bounds of possibility for a man to recognize the relative evil of his nature, but it is a rare and shattering experience for him to gaze into the face of absolute evil.'** (see par. 121 of *FREEDOM*)

[77] So we're starting to get a feel for how insecure, how distressed underneath, we are by our corrupted condition, and how we have been trying to live in denial of it all this time.

[78] The artist Francisco Goya's famous etching that he titled *The sleep of reason brings forth monsters* depicts how depressing **'reason[ing]'** or thinking about our corrupted condition has been. In it we see bats from hell tormenting the person's mind! (see par. 111 of *FREEDOM*)

Goya's *The sleep of reason brings forth monsters*, 1796-1797

[79] But while it's a very famous etching no one's ever really explained it. Yes, it is called *The sleep of reason brings forth monsters* but now we can *really* understand what he was drawing — it's a really powerful image of depression.

[80] And since pretty well any thinking brought us into contact with the issue of our horribly corrupted, seemingly utterly evil condition, virtually any thinking was unbearable — 'Oh, that's a lovely sunset,

ooohh, I wonder why I'm not lovable!'; 'Pass me the salt, have I told you about me, yes, okay, I'm so focused on me, I'm an ego maniac, get off my case!'

[81] I mean, any thinking, if you really sit down and look at it, will lead to confrontation with our corrupted condition, and when that happened you've had to flee from there.

Bacon's *Landscape*, 1978

[82] This is a painting the artist Francis Bacon once did of some grass, titled *Landscape*. He painted it from a photo he had of the natural world, which he said he then kept cutting up until he was

just left with a small piece of the photo, and this small remnant of grass was what he painted! I can't find the quote at the moment but he said something to the effect that this was all he could cope with in the photo of the natural landscape! [The actual quote is: **'I whittled it down and down until in the end there was just a little stretch of grass left which I enclosed in the box. And that really came about by trying to cut away, out of despair, the look of what is called a landscape...I wanted it to be much more artificial...I added that whole very intense surround of cobalt blue, which I felt made it look more completely artificial and unreal. I wanted that really strong blue to take all naturalism out of it'** (*Interviews with Francis Bacon*, David Sylvester, 1975 & 1980, pp.161-162 of 176).] By his own admission, Bacon was a very hurt soul, so it makes sense that the innocent, natural world was very confronting for him. I know people like hiding in alienated, **'completely artificial and unreal'** cities because they find the natural world too confronting. The more people, the more buildings, the more noise in our headphones, the more busy we keep ourselves, the better! So that's a further way we're constantly trying to keep at bay this confrontation with our corrupted condition.

[83] So, the comedian Rod Quantock wasn't joking when he said, **'Thinking can get you into terrible downwards spirals of doubt'**, nor was the Nobel Laureate Albert Camus when he said, **'Beginning to think is beginning to be undermined'**; nor was another Nobel Prize winner in Literature, Bertrand Russell, when he said, **'Many people would sooner die than think'**. And nor was another Nobel Laureate, the poet T.S. Eliot, when he wrote that **'human kind cannot bear very much reality'**. **'The sleep of reason'**, reasoning at a deep level, certainly has brought **'forth monsters'** as that picture by Goya depicts. (par. 121 of *FREEDOM*)

[84] I emphasise that the reason I've said HAS brought forth monsters is because, as I emphasised at the beginning of this talk, we now have the redeeming understanding of our corrupted condition that finally allows us to think about, and make sense of, anything and everything, which is absolutely **'the most wonderful of all gifts'**—as all our guests to the Sydney WTM Centre here today have been shouting to the world from every roof top they have been able to

climb onto. This is the *pre*-human-condition-understood situation that I'm describing. It is a 'has been situation', not an 'is situation'; we're now through that terror of not being able to understand our corrupted condition.

[85] So I'm not trying to scare everyone with the horrors of our condition by going through all this, <u>I'm explaining why there is an initial Deaf Effect when trying to read or hear about the human condition</u>—because it's only by understanding clearly why the Deaf Effect happens that everyone will be able to avoid being tricked and frustrated by it, and from there be better <u>able to persevere until they can get through the Deaf Effect</u> and receive **'the most wonderful of all gifts'** of being able to understand everything at last, and be forever free of the horror and agony of the human condition.

[86] So yes, we can understand why, up until now, we've had to have the attitude described in U2's 1992 song *Staring At The Sun*: **'It's been a long hot summer, let's get under cover, don't try too hard to think, don't think at all. I'm not the only one staring at the sun, afraid of what you'd find if you take a look inside. Not just deaf and dumb, I'm staring at the sun, not the only one who's happy to go blind.'** <u>The sun is clearly the confronting and exposing truth of our corrupted condition</u>.

Computer graphic by James Press © 2018 Fedmex Pty Ltd

[87] Yes, up until now virtually everyone had no choice during their adolescence but to 'resign' to hiding from **'the sun'** deep inside Plato's metaphorical dark cave [read about Plato's cave allegory in Video/F. Essay 11]. That's the only way we could cope. So that's where we have lived, in darkness—that's the real truth. This is why when it came to choosing a cover for my definitive presentation of this explanation, *FREEDOM*, we chose a vivid image of a sunrise to reflect the dawn of enlightenment that has arrived; the sun is finally coming up to drain away all the darkness from our lives!

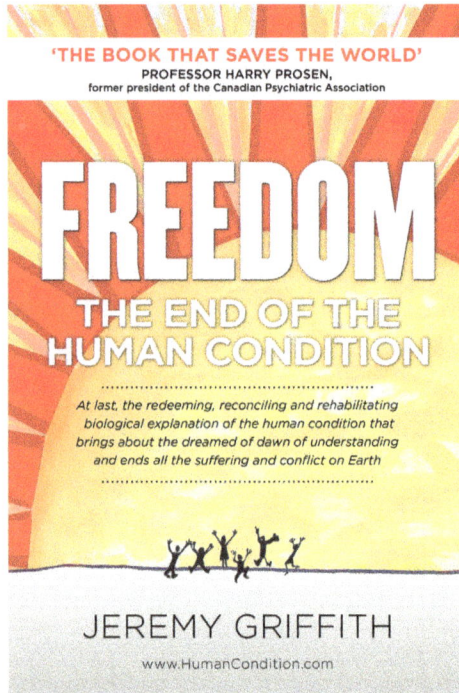

'THE BOOK THAT SAVES THE WORLD'
PROFESSOR HARRY PROSEN,
former president of the Canadian Psychiatric Association

FREEDOM
THE END OF THE HUMAN CONDITION

At last, the redeeming, reconciling and rehabilitating biological explanation of the human condition that brings about the dreamed of dawn of understanding and ends all the suffering and conflict on Earth

JEREMY GRIFFITH
www.HumanCondition.com

[88] I might mention in passing that some people have said about this cover that it looks more like a children's book than a serious, authoritative scientific presentation. My response to that is it is a

book for children in the sense that it contains the simple, innocent, truthful understanding that enables us to no longer have to fear the sun, but can live joyfully in its presence! I just love our cover but quite a few people have said that over the years.

[89] Returning to the important discussion of the transition that occurs during adolescence to living in denial of the human condition. All the pictures in F. Essay 30 about this process I have termed 'Resignation', such as those included below, make it very clear that trying to confront the subject of the human condition without understanding of it has been so unbearable for almost all adolescents that they had no choice but to eventually 'resign' to living in denial of the subject, a denial not undertaken lightly because it necessitates blocking out awareness of our all-loving and all-sensitive but unbearably condemning instinctive self or soul.

Stanisław Mikulski/AdobeStock;
mitarart/AdobeStock

[90] So these are photos illustrating adolescents, around 13 or 14 years old, who are still trying to wrestle with the imperfection of life around them and in themselves and the depression that this gives rise to. And the picture in the middle is particularly revealing; it's of a girl who's made her face into a snarling wolf, and the accompanying text says, **'It's not a phase Mom! This is who I really am!'** So Mum says, 'Look, you'll get over it, you'll get through that, it's just a phase'—all this adult resigned bullshit—but the girl's protesting, 'No, no, this is who I *really* am Mum, I'm a *really* bad person'.

[91] To illustrate this process with a rare, first-hand account, Alex Akritidis, who again, is here today, filmed this excellent description of going through Resignation.

[92] **Alex Akritidis**: '**My experience of Resignation, that I happen to remember, I recently wrote down: "After completing Year 9** [in secondary school] **when I was 15 years old, during the summer holidays I did absolutely nothing but stay in my room, in the darkness, preoccupied with my phone all day, every day. I never went out of my room; curtains down, lights off. I was so nutrient deficient and dehydrated, I would get up off my bed and see stars and feel a fainting sensation. I barely ate food, I barely spoke to my siblings, barely spoke to my parents; I would walk past everyone like I was in a different dimension. The alienation was so strong, I just felt like I didn't connect to anyone. Every time I walked out of my room, I felt like I walked into condemnation. Every person personified the reasons why I felt I wasn't good enough as a person. Those summer holidays was when I felt like I had become schizophrenic. I projected an imaginary scenario each day of around four people on each side of my room that I received some sort of validation from. I couldn't bear the agony anymore of the exposure of the imperfections within me and within the world.**

[93] So the following year, after the summer holidays I started Year 10 as if I had hopped into a completely new body. I was numb and felt like a new person, as if I had shed my soul from my body. I was completely absorbed by my masks and coping strategies. I was so embedded into denial that I could no longer acknowledge any imperfections within me, and I fully subscribed to pseudo idealism and my victim mentality to extract reinforcement from my surroundings.'"

[94] So at that point I just became soul dead, a walking zombie.'

[95] **Jeremy**: Wow, well done, Alex. Good one, mate. That's powerful and will help a lot of people. So you can see how he's trying to face down the human condition in the world around him and in himself, and it's so unbearably depressing that he's crippled, he's totally crippled. Young adolescents at some point have had to just give up trying to wrestle with the human condition and leave it behind and just get on with life, and at that point they become an artificial, pseudo, seeker of power, fame, fortune and glory. They get on with life and can't remember any of that terrible wrestling match. A lot of people can't even remember going through Resignation, so Alex did well to remember that; I mean he's pretty young and so still within memory of going through it, but again, his recollection gives us some idea of why the Deaf Effect happens because we're taken back to that interface with this horrific issue of our corrupted condition. *This* time, however, we're beautifully defended and we can be free of that condemnation, but we have this historic fear as a result of that initial depression, and so we're still fearful that we're being taken back to that dark corner.

[96] And this practice of denial and its blocking out of our soul is now so deeply entrenched in most people that, as the great Scottish psychiatrist R.D. Laing said, **'there is a veil which is more like fifty feet of solid concrete'** between us and **'our true selves'** or soul (see source shortly in par. 129). Fifty feet of solid concrete is how much we have blocked out the truth about our corrupted condition, so that's a good measure of how extreme the Deaf Effect is going to be!

[97] The quote most frequently referred to in all of Sir Laurens van der Post's books is this line from Gerard Manley Hopkins's poem *No Worst, There Is None*: '**O the mind, mind has mountains; cliffs of fall, frightful, sheer, no-man-fathomed**'. Yes, there was 'none' 'worst' than the 'frightful' suicidal depression that thinking about the human condition could cause while it was still to be 'fathomed' or understood. Thank goodness the human race doesn't have to face any more of this terrible, terrible depression.

[98] Yes, the problem has been that, as the science writer Roger Lewin once said, trying to '**illuminate the phenomena of consciousness**' is '**a tough challenge…perhaps the toughest of all**' — 'consciousness' being the evasive code word mechanistic scientists have used for the issue of our corrupted human condition—rightly so because 'consciousness' is at the heart of the issue of why we became so messed up, and what causes us to be worried about being so messed up. Consciousness appears to be the culprit! (see par. 624 of *FREEDOM*) So it's a good code word to use—but that's how evasive we are. Intellectuals just talk about consciousness; they say 'consciousness' and everyone thinks, 'Yeah, I know what you're on about, don't go any further than that' and that's what's been going on—all this rubbish.

[99] Which all explains why my wonderfully honest professor of biology when I was at Sydney University, the Templeton Prize winner Charles Birch, wasn't exaggerating the 'failure trapped' blindness of human-condition-avoiding, mechanistic science when he said that '[mechanistic] **science can't deal with subjectivity…what we were all taught in universities is pretty much a dead end**' (well, it's not getting into anything is it, it's just bullshit); and '**the traditional framework of thinking in science is not adequate for solving the really hard problems**' (yes, they won't go anywhere near what the real issue is of the human condition); and '**Biology has not made any real advance since Darwin**'; and '**Biology right now awaits its Einstein in the realm of consciousness studies.**' (par. 625 of *FREEDOM*)

[100] Yes, what has been desperately needed is someone innocent enough to look into the subject of the human condition and by so

doing find the redeeming and healing understanding of it is — which is what I've done — and by the way, this doesn't make me any more special or worthy or deserving than anyone else because exceptional innocence is just one of the inevitable positions in the spectrum of alienation that the human race's heroic search for knowledge unavoidably led to. A fundamental insight that understanding of the human condition gives us is the equal goodness and worthiness of all humans. We humans have *all* been involved in a great and necessary battle, so we have inevitably *all* been variously knocked around in that great battle, but we are *all* heroes — even people like me who were lucky enough to not have been involved in the thick of the battle during their infancy and childhood are heroes. Everyone is actually a bigger hero than me, and it's true, but I can still claim to be a hero, so there you go! [Audience laughter]. What I have just said is the truth, I am less a hero than anyone in the room, which is why last night, Lucas [Machlein], I didn't want you thanking me for any damn thing.

[101] **Annie Williams**: He said, 'How can I ever thank you? How can we ever thank you.'

[102] **Jeremy**: Oh shit, I don't want to go around this loop, I've just been through it [audience laughter]. It's true, it's amazing what I've done but you've got to understand, I didn't set out to be a legend or anything, I just did what everyone's trying to do — make sense of my life — and since I was coming from a very innocent, secure situation, that's the way I got here. Like those quotes from Tony [Gowing] and Sam [Belfield] evidence, people tried but couldn't make sense of anything, there was no meaning, there was no structure, everything was just a mess and everyone knows that and everyone tries to sort life out but can't get anywhere; they can't get to the bottom of it at all, they can't even *begin* to know where to start tackling it. So it's an absolute astonishment that somebody did unscramble it. But the way I did it was that I just did what they

and everyone else was doing—trying to think and make sense of life, but, again, I'm coming from an incredibly innocent, secure position so I'm thinking truthfully—and I was struggling because, as I'll talk about a bit later, the world was so damn mad and I could not understand it *at all*. It was a huge problem for me, and then my mother gave me this book by Sir Laurens van der Post; it was *Venture to the Interior* because I remember the picture of the zebras running across the cover.

[103] You see, van der Post is an honest thinker, he talks about the relative innocence of the Bushman [of the Kalahari], and he was revealing truths, paragraph after paragraph, and I was thinking, 'This guy is telling me I'm not mad after all.' So that saved my life. I was then untouchable; no one was ever going to shake me and they never have, because they're bullshitting and I'm not, and I'm hanging onto what I know is true and nothing else matters, and I have no interest in that other deluded and dishonest world, and that's how I got to the explanation of the human condition. I just

kept thinking truthfully from that basis, and so it is miraculous when you stand back at this point, but I'm just a human trying to make sense of life like everybody else, but I'm just coming from that very innocent position. So, again, I didn't set out to be a legend or any damn thing. And I'm not—as I fully explain, I'm just at one end of the spectrum of alienation as a result of this heroic battle. If there's a great battle and you've got some people in the middle of it cut to shreds, dying of depression from extreme alienation, they're the most heroic to have endured that. I was someone who missed out on being exposed to the human condition during my infancy and childhood; I was sitting under a tree scratching my ear when everyone else was getting their heads chopped off. And then I come out and say, 'I've solved the human condition', but they're the legends. Honestly, this understanding sorts it all out, it makes sense, it gets rid of all that shit, all that elitism and...

[104] **Annie**: It demystifies the whole thing.

[105] **Jeremy**: Yeah.

[106] **Annie**: And like Franklin Mukakanga [WTM Zambia Centre] said, it's just the power of a soul still alive, how you've been able to solve the human condition.

[107] **Jeremy**: Yeah, I know it's pretty amazing [Annie laughs]. I walk around saying, 'That's not a bad effort!'; I think I might get on this ego bandwagon [audience laughs]. I make jokes with Tony [Gowing] about it; you know, when he compliments me, I will say, 'Come on, say that again!' [Laughter.] I was playing that game with you Lucas, yesterday; we were talking about Alex's ability to sing, and I said to Lucas, 'So where's your fucking song mate?' [Laughter.] But if you watch what I'm doing with people, even when I'm apparently joking around, all the time I'm trying to take the load off their shoulders, trying to make their load lighter and help them. And Lucas is a really sensitive soul and I'm so pleased

you're out here mate because this information and project will look after you like nothing else. Sorry, I better get back on the job.

- - - - - - - - - - - - - - - - -

[108] So the human condition has been solved, but there is still one massive impasse to overcome—which is that this historic fear we humans carry of the issue of the human condition blocks the ability to hear and read about the human condition and discover that it's finally been made safe to confront, and from there, discover **'the most wonderful of all gifts'** of being able to understand themselves and the world, and completely move on from the human condition.

[109] This Deaf Effect problem is VERY real. For example, there is a Swiss bloke who's been promoting this understanding of the human condition for years (but is not affiliated with us in any way). To try to break up people's historic denial he puts them on mushrooms or some drug so they can start to hear this (a practice we don't condone). And he's got all these meditative techniques to help people to access this understanding; that's his way of solving the Deaf Effect. He understands the struggle, giving this accurate description of his experience with it: **'the root problem is the guilt, which is buried so deep, it's so hard to get there, how much courage that needs…when the question pops up of if they're good or evil, the possibility that they are actually evil, they shy away, they try everything to avoid going to that point'.**

[110] This is similar to what Ari said: 'As soon as I started reading that I'm addicted to power, fame, fortune and glory, or whatever, or I'm egocentric or I'm selfish, I can't accept that' because denial of his upset state has been the basis of his whole existence. Ari—like every bloke—has built this massive defensive castle, and he gets a few wins and he's been made head of the fire brigade or whatever other win you can find. And so you swan around on the basis of these superficial and artificial wins and keep the insecurity and

depression at bay. As Tony Gowing's always teaching us, every day, every moment is focused on trying to maximise the positives and minimise the negatives, and by posturing and carrying on—it drives everything we do; for example, 'I've got a nice puffy blue jacket, don't I look great, aren't I a winner', etc, etc. It's all this masquerading, and it's all because we're so deeply insecure. And so as Ari said, 'As soon as I started reading that I'm supposedly competitive, aggressive and selfish, I said, 'That's not me, I don't relate to that.' But the truth is, as *FREEDOM* goes on to explain, we're 2 million years corrupted, let alone just a little bit! So it's true, people switch off. And so while I haven't been in touch with that Swiss guy for a long time and don't know how he's going, what he does to get people to overcome the Deaf Effect is give them my books, and then get them to meditate, take certain mushrooms, stand on their head, and God knows whatever else, to break up their historic block, which is *incredibly* strong, and then they can hear this.

[111] So yes, the one final challenge for the human race now is to have enough people overcome the Deaf Effect so that there is a critical mass of appreciative people to show everyone else that it is actually safe now to confront the human condition.

- - - - - - - - - - - - - - - - -

[112] At this point I want to mention that there is a small category of people who don't particularly suffer from the Deaf Effect. They are what we in the WTM call 'Ships at Sea', people who, for various reasons—in some rare cases because they are relatively innocent and so haven't had to resign—bravely refused to 'pull into port' as it were and resign to living in denial when the 'storms out at sea' of unbearable self-confrontation occurred during their adolescence.

© 2022 Fedmex Pty Ltd

[113] Lucas here is a 'ship at sea', so I will play his description of how he couldn't understand the resigned world and refused to buy into it:

[114] **Lucas Machlein**: 'Before I encountered the information I was actually really stuck in myself and I wasn't even trying to look at the world anymore. I wasn't watching the news, and I was pretty concerned just about my own human condition and trying to hide it from people, trying not to show I was depressed, not looking at people, taking distance from my family, travelling to another country—and always being on the run, always changing groups of friends and changing work. I was just trying to wrestle with and understand what was going on in my head and that's when I found this information and that was really timely, because I was really starting to wonder whether I was just going mad; I was starting to ask people that. Thankfully it seems not! And now I feel healed. I feel I understand myself. I understand my life and I can understand other people's lives. I can understand my family, the people I grew up with or even people I just met. I can tell what they are trying to do, the strategies they use, how they cope with life, all of which used to distress me enormously. It's pretty exciting. It's a wonderful tool, but it's more than a tool. It's not just a way to understand and "screw around" people, it's a way

to really understand people to the core and be able to love and appreciate them for who they are and for what they are doing.'

[115] **Jeremy**: Good one Lucas, well done mate, good effort. That's pretty special. I've only spent half an hour with Lucas but I can tell that he's a sensitive person. Most people, once they resign, as Alex was saying, get on with life and it's all about escapism and power and glory, and distraction and the next party and getting drunk and whatever. It's just 'rip and roar'. This other pre-resigned state, where you're 'back there' stranded trying to make sense of the world, is a totally different existence—so people in the midst of Resignation *are* totally different to those who are post-Resignation, which Alex described so well.

[116] Olof Österman [pictured overleaf] is the founder of the WTM Sweden Centre, and he also visited us in Australia, and he's a similar 'ship at sea'. Because he didn't give up on looking at the world honestly, he's still very, very distressed and traumatised by it. But 'ships at sea' who discover this understanding cannot believe the relief they get, there's just no Deaf Effect. Another example of this is a 16-year-old girl who said, 'I read one of your books all in one night, and it was just such a relief.' [This is what this girl, Lisa Tassone, actually wrote: **'Before stumbling upon *Free: The End Of The Human Condition* that was discreetly shoved in the back of the philosophy section, I was at the end of my road. I had experienced a year of complete and utter pain, confusion, anger and frustration. When I finally took the plunge to seek medical help (as I was suicidal), I was diagnosed with severe depression and put on medication. After reading your book (which I stayed up till 2am reading, I just couldn't put it down), I have been one of the fastest recovering depressants around. No wonder why. If everyone knew your insights, so much would be resolved. The purpose of this letter is to thank you for your courage in publishing your sure-to-be controversial work, and for basically recovering and saving this 16 year old. Not only is your work the absolute truth and has restored my faith in humanity, it has given me inspiration to help others. I**

**may seem young to know what I'm talking about but, well, I do. I have tested
all your work and others and yours always held up'** (4 Oct. 1999).]

[117] And so there's an absolute world of difference between the
pre-resigned state and the post-resigned state. Once you get to the
post-resigned state, once you resign, then obviously this whole issue
of our corrupted state is something you've blocked out, otherwise
you're constantly living in a fearful state of depression, like Lucas
described—how he was just running away from everything and
couldn't cope; and Alex was just locked in his room, and so on.
So to be able to make sense of all that brings the most incredible
relief.

Stefan Rössler (left) and Olof Österman (right) in Australia, 2019

[118] And it's these 'ships at sea' who haven't resigned to living
in denial of the human condition who can most help get the support of this world-saving understanding going. They—and anyone
else who has been able to either avoid or get through the Deaf
Effect—have to lead the world home to freedom from the human
condition.

[119] You will notice if you read the ecstatic responses on the slider at the top of our homepage how quite a few of the people published there are 'ships at sea' because it's clear in their comments that they don't suffer from the Deaf Effect and can immediately take in or 'hear' and have become excited about the explanation of the human condition—for example, these are two from the last few days: **'I always wondered why I thought and acted differently. This is my confirmation beautifully explained. I am now truly free'**, and **'Just what I wanted to hear to understand why I am not abnormal.'** And I think Stefan Rössler [WTM Austria, shown above] said he thought he was an alien [laughing]. There are a few 'ships at sea' who've said they think they must be aliens. They get so distressed, their thinking is so different that they finally decide that 'Look, I must be an alien from another planet, or everyone else must be. I have no relationship to all these people.' Like Alex's pre-resigned self wouldn't have been able to relate to his resigned self that was just on a bender of distraction, escapism and delusion, joining the Environmental or Climatism or Woke movements to make himself feel good, the **'pseudo idealistic' 'coping strategies'** that Alex said he had to **'fully subscribe to'** to escape feeling **'condemnation'**, and so his pre-resigned and resigned self are totally different people.

- - - - - - - - - - - - - - - - -

[120] So up until now, only young adolescents who are yet to resign, and rare exceptionally honest thinkers like Sir Laurens van der Post and R.D. Laing, and rare exceptionally brave artists like Bacon, Goya and Munch, could look at the human condition with any degree of honesty.

[121] The following examples of such people serve to illustrate just how alienated the human race has become and therefore how precious **'the most wonderful of all gifts'** of being able to understand the human condition now is.

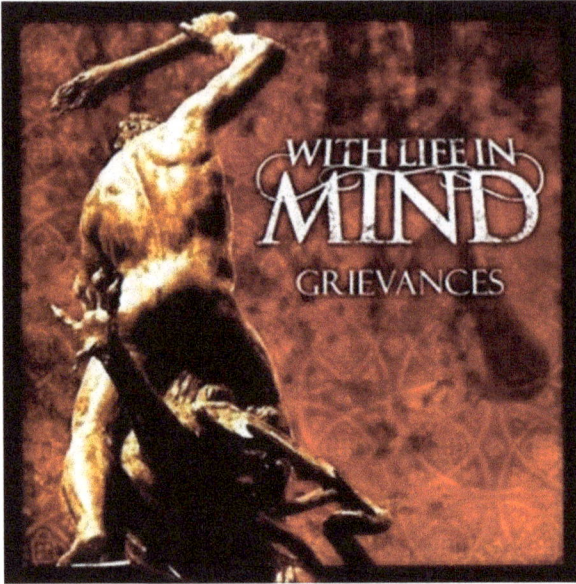

[122] An incredible <u>example of a pre-resigned honest mind</u> can be found in the lyrics of the American heavy metal band With Life In Mind's album *Grievances*, which were written by one of the band's members when he was still a young teenager. And look at the cover [above], the world is mad, you know—here's this person screaming, getting beaten up. With Life In Mind, I mean, that should be With The Human Condition In Mind! So these are the lyrics, which are just unbelievably free of any resigned thinking. If you want a true description of our world, this is it: **'It scares me to death to think of what I have become…I feel so lost in this world'**, **'Our innocence is lost'**, **'I <u>scream</u> to the sky but my words get lost along the way. I can't express all the hate that's led me here and all the filth that swallows us whole. I don't want to be part of all this insanity. Famine and death. Pestilence and war. A world shrouded in darkness… Fear is driven into our minds everywhere we look'**, **'Trying so hard for a life with such little purpose…Lost in oblivion'**, **'Everything you've been told has been a lie…We've all been asleep since the beginning of time. Why are we so scared to use our minds?'**, **'Keep pretending; soon enough things will crumble to the ground…If they could only see the truth they would coil in disgust'**, **'How**

do we save ourselves from this misery…So desperate for the answers…We're straining on the last bit of hope we have left. <u>No one hears our cries. And no one sees us screaming</u>', 'This is the end.' Wow, that's pretty good isn't it? What do you think of that, Alex? That's not a bad description is it?

Munch's *The Scream*, 1893

[123] So this next picture is Edvard Munch's *The Scream*, which is an incredible example of the honesty some <u>artists</u> are capable of, because he dared to depict the 'scream' that With Life In Mind said 'no on hears' or 'sees'. This is a really powerful and very famous picture—everyone has it, all the teenagers have it pinned up in their rooms, but like Goya's *The sleep of reason brings forth monsters*, no

one's ever really explained it, not like we can now. It's really powerful, because see the two people in the background of the painting
who are promenading down this pier on the edge of the sea—these
two people are like normal humans, 'normal' being totally resigned
to pretending everything is fine and as it should be. So they're
swanning along the pier saying, 'That's a lovely sunset, should we
go down and have an ice cream?', whatever the hell, and yet this
person is screaming his head off and the whole environment is just
resonating with the horror of it. So these two people, contrasted
with that person screaming, is a really, really powerful presentation.
The bravery of Munch daring to confront the human condition was
made clear when he said that at one period in his life, '**My condition
was verging on madness—it was touch and go.**' (see F. Essay 44)

[124] So to be an artist and try to cut your way through to the truth,
a window into the truth, like Vincent van Gogh did, was incredibly
brave; he painted light, he taught us to see light. We couldn't see
light until Van Gogh painted it—it's true! Because he was just so
honest, and he just kept making himself more and more honest in
how he saw the world, until he could see its true illuminated beauty.
So he painted these amazing pictures and when you look at them
you can really see light for the first time, the brightness of it.

Van Gogh's *The Sower*, 1888   Van Gogh's *Three Sunflowers*, 1888

[125] He could paint anything, like even these sunflowers in a vase, and suddenly it came alive. But he cut off his ear and went mad and suicided, poor bugger. To be an artist, to try to dig into the truth, through whatever form of art you were pursuing, was a tortuous existence—but confronting the truth no longer is because we can now understand the whole damn journey.

[126] Now this is a really good example! You'll like this one. In the case of the artist Francisco Goya, we've already seen his courageous honesty in *The sleep of reason brings forth monsters*, but he also did two contrasting paintings that are incredibly revealing of the human condition, which I include in Freedom Essay 44. As the esteemed art critic Robert Hughes described them: **'There are two paintings of the same subject...**[They are of] **a big religious festival, that of St. Isidro. On that day thousands of citizens, in their Sunday best, converged on a pilgrimage chapel outside Madrid and had a picnic.'** In the first representation titled **'***St. Isidro's Meadow***'** [below], Hughes said, **'the girls are in their white parasols, the men in their finery, the scene is of social pleasure and jollity'**.

Goya's *St. Isidro's Meadow* (detail), 1788

[127] Then, according to Hughes, '**Thirty years later Goya returned to
the same theme. In this picture** [below, titled]**...*The Pilgrimage of St. Isidro*,
instead of these happy fashionable well-dressed young people, you have this
horrible snake of...dark figures...like demons crawling across an ash heap. The
faces are...of madmen and hysterics...The whole picture is deeply threatening'**.

Goya's *The Pilgrimage of St. Isidro* (detail), 1821-1823

[128] Goya clearly knew humanity was living a completely fraudulent,
escapist, deluded existence. Accompanying one of his etchings
he even wrote that '**The world is a masquerade. Looks, dress and voice,
everything is only pretension. Everyone wants to appear to be what he is
not. Everyone is deceiving, and no one ever knows himself**' *(Capricho 6)*. So
Goya was a brave man. And this bravery stands alongside that of
the pre-resigned kid from With Life In Mind who wrote that band's
excruciatingly honest lyrics.

Peter Davis 1967

The irreverent R.D. Laing (right) in 1967, addressing
the 'Dialectics of Liberation' conference

[129] He has already been introduced in this presentation, and part of the following quote cited, but the exceptionally honest R.D. Laing's description of our human condition is without peer in articulating the alienation from our soul that Goya painted and With Life in Mind screamed out to be heard, and so is worthy of full inclusion here: **'Our alienation goes to the roots. The realization of this is the essential springboard for any serious reflection on any aspect of present inter-human life...We are born into a world where alienation awaits us. We are potentially men, but are in an alienated state** [p.12 of 156] **...the *ordinary* person is a shrivelled, desiccated fragment of what a person can be. As adults, we have forgotten most of our childhood, not only its contents but its flavour; as men of the world, we hardly know of the existence of the inner world** [p.22] **...The condition of alienation, of being asleep, of being unconscious, of being out of one's mind, is the condition of the normal man** [p.24] **...between *us* and It [our true selves or soul] there is a veil which is more like fifty feet of solid concrete. *Deus absconditus*. Or**

we have absconded [p.118] …The outer divorced from any illumination from the inner is in a state of darkness. We are in an age of darkness. The state of outer darkness is a state of sin—i.e. alienation or estrangement from the inner light [p.116] …We are all murderers and prostitutes…We are bemused and crazed creatures, strangers to our true selves, to one another' [pp.11-12] (*The Politics of Experience* and *The Bird of Paradise*, 1967). 'We are dead, but think we are alive. We are asleep, but think we are awake. We are dreaming, but take our dreams to be reality. We are the halt, lame, blind, deaf, the sick. But we are doubly unconscious. We are *so* ill that we no longer feel ill, as in many terminal illnesses. We are mad, but have no insight [into the fact of our madness]' (*Self and Others*, 1961, p.38 of 192). 'We are so out of touch with this realm [where the issue of the human condition lies] that many people can now argue seriously that it does not exist' (*The Politics of Experience* and *The Bird of Paradise*, p.105).

[130] Yes, I was watching someone going along picking up litter from a footpath and he was so enthusiastic and seemingly feeling fulfilled and meaningful by doing it that I felt sure he was believing that everything in the world is relatively ordered and in good shape, and he just needed to clean it up a bit to make it perfect. It was like he had no idea there is 2 million years of psychotic alienation from our true self or soul in us humans, in fact that we're fast approaching terminal alienation and the extinction of our species! Just as Laing said, 'We are the halt, lame, blind, deaf, the sick. But we are doubly unconscious. We are *so* ill that we no longer feel ill, as in many terminal illnesses. We are mad, but have no insight [into the fact of our madness]'.

[131] Again we can see how extremely unaware resigned humans have been to the horrifically corrupted state of the human condition and therefore how hearing or reading about the human condition initially causes them to experience a complete Deaf Effect.

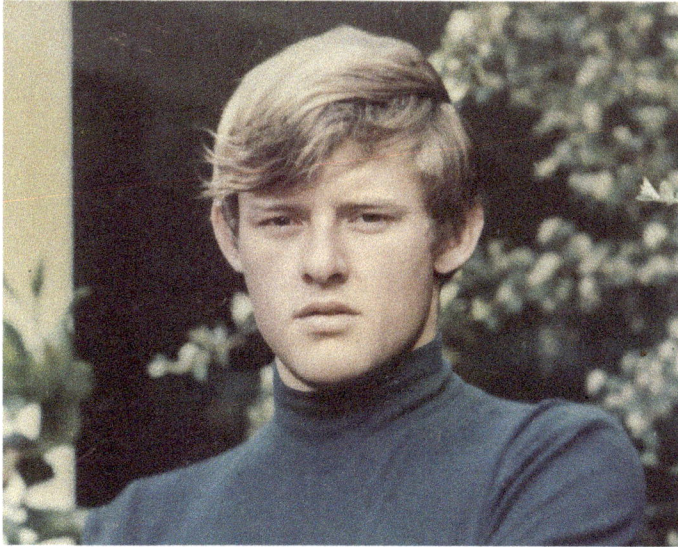

Jeremy in 1967, aged 22

[132] When I was growing up, being so innocent I was, as mentioned earlier, completely and utterly bewildered by the world of adults because they were carrying on as if everything was fine and I could see that it wasn't, so it was all an incredible agony for me, so much so I almost had to find this understanding of the human condition to save my life. Many times I would run away through the night from gatherings because I thought everyone was so fake and artificial and dishonest. On one occasion in my 20s I stood on a chair and in a loud voice told all the celebrities in this vast room that they were all frauds—that didn't go down well! Anyway, that's all history now, as it is for the whole human race, because the whole human race is coming home at last to soundness and sanity.

[133] So that brings me to the end of this talk that explains why, to quote a former prime minister of the United Kingdom, Benjamin Disraeli, our species has been **'stranded halfway between ape and angel'**, and how we can now all become **'angels'** again, people living happy and free of the human condition—which my soul has joyfully drawn here.

- - - - - - - - - - - - - - - -

[134] **Annie Williams**: I found that talk so astonishing, just extraordinary. For the soul-alive to speak to the soul-dead—the resigned state is necessarily and so heroically, like you've said, a state of darkness and untouchability because you're so defensive against anything that takes you near the human condition, and if anything does, you just get absolutely furious. And it's going to take the power of that soul, that incredible empathy and sensitivity and compassion, to

reach over there and tell us that we were once that sensitive and happy and joyful and full of love that the mysteries of the physical world didn't even bother us. I mean, that just hit me in the chest. And you say this is like a weapon against the Deaf Effect and I just pray that will reach everyone, because like you say, you've got to buy some time for their brain to be tolerant enough to actually hear the defence of that blackness. And that's what was so astonishing about that talk: you took us all back right to the beginning, where we all started, when we were all sensitive and beautiful and loving. And I just feel that really could reach us. Like van der Post wrote, the men whose hearts have become stone [see F. Essay 51], and they had to become stone—my heart is stone! And when I listen to you, that softens and it lets it in and I feel loved and defended and so inspired. So I just think that was the most astonishing talk, Jeremy. Thank you.

© 2022 Fedmex Pty Ltd

- - - - - - - - - - - - - - - - - -

Addendum – The most wonderful of all gifts, by Refentse Molosiwa

[135] This relieved and inspired response from Refentse Molosiwa, who is a young South African author who studied philosophy at the University of Pretoria, shows the life-transforming and world-saving benefits for everyone of <u>overcoming the Deaf Effect with perseverance</u>.

[136] As Refentse summarises about the significance of Jeremy's synthesis, **'I honestly don't know how we can go forward in the world without Jeremy's information.'**

Watch the video of this presentation at
www.HumanCondition.com/Refentse
OR
Scan code to view:

Refentse Molosiwa in Mahikeng, South Africa, August 2022

[137] **Tony Miall** [Sydney WTM founding member]: Tell me a bit about yourself. How long ago did you come across the information?

[138] **Refentse Molosiwa**: About maybe a month and a half ago. I was on YouTube ironically searching a video on existentialism and one of the adverts was *THE Interview*. I'd seen the advert before one time, but I ignored it, I didn't click on the link the first time. Then the second time I was like, 'hey, maybe there's something here' because the way Craig Conway said 'Stop what you're doing, this is a very important interview, you have to check out.' I was like 'Okay, let me give it some attention', because naturally I'm a very curious person and I like reading.

[139] And when I watched *THE Interview* my mind was blown away instantly. I was shocked, and luckily for me, how I was raised [relatively nurtured with love and sheltered from upset] meant that I didn't suffer from the Deaf Effect when I immediately read the books. Because after watching *THE Interview*, all I did was download all of Jeremy Griffith's books. I started with *Death By Dogma* and then I went straight to [Jeremy's definitive presentation] *FREEDOM: The End Of The Human Condition*. *Death by Dogma* was a good introduction. It was nice and short—straight to the point. Nice and punchy, I liked it. I highlighted almost the whole book because it was just so shocking—specifically, because ever since COVID started I was not very aware of liberalism and everything that comes with it, or the dichotomy between left and right—so when *Death by Dogma* was talking about liberalism and the 'feel good' nature that the liberal way of life is trying to impose, it made absolute sense because that was one of the biggest things I felt about COVID. There's this 'feel good' factor that is trying to be imposed on the world; that we must all follow this 'everything is good', virtue signalling type of situation. And that really hit me hard and I was like 'man, this actually makes so much sense to see what's really going on'; and how the Right is so much about pushing individualism and 'climbing the mountain [finding

understanding]', if I may put it that way. And then the Left is the very opposite; it meshed so well.

[140] And then I went on to *FREEDOM*, which was just 'wow'. Wow, wow, wow, wow, wow. Because it went into so much more depth with everything, like everything, everything, everything. And the deep things I'd been looking for the answers to for so, so, so long. It's unbelievable. And it makes sense when Jeremy says we have been on this journey for 2 million years. Because just in my own life I've been looking for these answers and it feels like it's been an entire million years in just one lifetime!

[141] So I just kept going, kept going, and when I found *FREEDOM*, wow. I really can't describe how earth-shattering it was, if I'm to be honest. Especially when he was speaking about the bonobos [see chapter 5 of *FREEDOM* or F. Essay 21], and how we got our consciousness [see chapter 7 of *FREEDOM* or F. Essay 24].

[142] Every time I tried to 'resign' in my life and give up the search for knowledge it was very difficult [see chapter 2:2 of *FREEDOM* or F. Essay 30 on 'Resignation']. It was very, very difficult. I became very obsessed with trying to understand why humans behave the way we do. I was always questioning why do we do this? How is this possible? How can we be so mean to each other? It just didn't make sense to be honest. *It just did not make sense.* And honestly, Jeremy's book, out of I don't know how many books I've read or pieces of information I've looked into trying to understand human nature, nothing comes close to Jeremy's work. Absolutely nothing. Absolutely nothing. It's true when they say Jeremy is a prophet in that book. It sounds like an exaggeration, but it's absolutely true. It's absolutely true. And when he broke down the meaning of God, oh my goodness gracious, Integrative Meaning—oh! [see chapter 4 of *FREEDOM* or F. Essay 23.] That was like being hit, like running into a wall, and it was just like, wow! I really have no words to describe the profundity of the understanding. It made my life make sense.

[143] It made my life make absolute sense. Specifically the process of Resignation, which I think everyone can relate to. That's one of the most easy things to relate to because I think almost everyone goes through that. That was one thing that I thought, 'Okay, cool'; I genuinely resonated with that so much. I can really understand now why people give up, because a lot of my own journey was very difficult. It was very, very difficult to keep going. I'm just glad I had the strength to go on. I can only imagine how hard it must have been for Jeremy to actually have been pushing all this knowledge for so many years alone. Like, wow. That takes an immense amount of strength for him to have even released a book. It's just incredible; it's incredible.

[144] In my life at least, I was fortunate that growing up I was surrounded by many books. So I remember the first book I read was *The Child Called It* by Dave Pelzer [about child abuse]. And I would say that was probably the first time in my life I really encountered the upset of the world through that book in particular. And from then on, I was just so curious to understand why, especially in the context of that book — which was a true story — I was so shocked. How could a mother do this to her child? And from then on, I think that book made me question more than anything else, why do people do what they do? It doesn't make sense for a mother to be so terrible to her own child. It was just so shocking and it made me keep asking 'Why?'

[145] The metaphor of the sun [see F. Essay 45] is so tremendous and it makes sense now because when I was in my first year at University of Pretoria, I did philosophy and that's when I came across Plato's 'Allegory of the Cave' for first time — but I really didn't understand it to the depth that Jeremy describes it [see F. Essay 11]. Specifically, that once you can understand Resignation, it's incredible because once you at least are able to absorb Jeremy's information it's shocking how resigned humans are and how hard the human condition has hit

us for so long. And man, we've been heroes inasmuch as we have done wrong in many aspects but for the fact that we're still here, still fighting—wow, humans have been so heroic. We didn't give up. You know, like Jeremy didn't give up, and the few people out there that didn't give up and that in itself is something that should truly, truly be celebrated, because I honestly don't know how we can go forward in the world without Jeremy's information. No, I don't see a way, it's so important. I honestly don't know of any other way—*Jeremy's book is the greatest book ever written in my opinion.* I can't think of any greater piece of information out there that is so healing and reconciling because that word 'reconciling' is very big because all aspects are involved in the human condition.

[146] I don't think there's a part of life that the human condition doesn't affect, especially when it comes to human relations and how we behave. From relationships to everything. Specifically, another reason why this understanding hit me so hard was I had broken up with my girlfriend just before COVID, which made my search even more like 'Why is this happening?', 'Why is it so difficult for men and women to connect?', 'Why are relationships so difficult?' [see chapter 8:11B of *FREEDOM* or F. Essay 26 and 27 on men and women.] All this stuff. So when Jeremy explained it in the books as well, I was like, 'My goodness gracious, how is it possible for one man to have all these answers? This is just so crazy and it makes so much sense!' That's what makes it even more incredible. Wow! There's no dogma, there's no belief system, there's no 'You must believe this', or something like that. No, it's just 'Use your mind.' You have to just think. And I was like, 'Wow, this makes absolute sense, this makes absolute sense.' I've never looked back ever since, to be honest. I don't know how to put it in words. It's just profound.

[147] I was afraid that if I didn't continue searching for knowledge and just resigned [to living in denial of the whole issue of the human condition] that I was going to commit suicide. The moment

I stopped looking for the answer I know deep down that I'm going to be living a tremendously superficial life. I wouldn't know how to go on. Without answers, there's no way I'll be able to go on just living the 9 to 5, putting my head down and just giving up the search; I couldn't, I couldn't deal with it. That to me was more scary than searching for the knowledge, because at least there was some hope with the knowledge. As long as I keep going, there's something out there. Because I remember some quote by Buddha that always kept me going was **'The only two mistakes you can make on the path of truth: one is not starting; two is not going the whole way.'** And since I started, I cannot stop. I have to go the whole way and even if I die trying it's better than resigning and giving up the whole search because to me giving up was death. Giving up was suicidal in my mind.

[148] The healing understanding that Jeremy Griffith has given us is the only way. And I was like, there *has to be* some sort of way because there was no other explanation because nothing made sense, nothing made sense. And all the different types of philosophies— New Age; I went into economics, education, conspiracies; I read people like David Icke. I went down the rabbit hole of almost all the things that you can find. Politics, religion. I was once deep into Christianity, but I got out of that. I went into other Eastern religions, I got out of that. So wow, I truly cannot really sum up how powerful Jeremy's work is. As an African specifically, when he talked about Integrative Meaning [see chapter 4 of *FREEDOM* or F. Essay 23], the selflessness, living together, it clicked immediately because here in South Africa we have the term 'Ubuntu' ['humanity']. And in one of the other videos that I watched, there's another guy from Eastern Cape called Reginald Khotshobe and he spoke about the same thing [Reginald is the founder of the WTM Eastern Cape Centre]. And I was thinking this makes a lot of sense because when I was reading about Integrative Meaning and it was talking about the selfless love, the word 'Ubuntu' came to mind immediately, and I was like, 'wow, this makes a lot of sense'. And also having grown up in the village

with my grandmother, my cousins, that 'Ubuntu' lifestyle was what we grew up in. All of us were together. We all lived sharing things; it was very communal. So it made a lot of sense, it's just natural. This Integrative Meaning/idea of God literally makes a lot of sense. Even my life, this makes a lot of sense. It makes rational sense. It's logical, it makes sense.

[149]One of the things that I also experienced as a black South African living in South Africa, given our history of apartheid, was that as I was going down the rabbit hole of trying to search for the truth, and especially if you're going to go down the path of studying African history, you invariably develop some sort of animosity toward white people. But the wonderful thing about Jeremy's work is it was just so healing, so reconciling that every form of animosity I ever had to white people or anything like that, it was all wiped away in an instant! Everything made sense. Racism actually made sense [see chapter 8:16E of *FREEDOM* or F. Essay 28]. Okay, this is actually what it is: different ethnic groups in the whole human race have different levels of psychological upset, and ethnic groups with less psychological upset may be antagonistic towards another ethnic group with a higher level of psychological upset. That was one of the most reconciling explanations I've ever heard in my life, to the point where I don't see colour anymore—not in the sense that we're not different, but we're suffering from one thing, we are suffering from the human condition. This is actually what's going on. And once you can understand that, then the borders we have in terms of not even just physical borders, but colour borders or prejudicial borders—all these sort of artificial borders we've made that stop us from connecting with one another, all those borders—literally disappear through that understanding.

[150]Which is so incredible because I remember reading this Indian philosopher called J. Krishnamurti—probably in my mind he is right up there with Plato—he didn't come up with the [Cave]

Allegory but he's right up there. And one thing he always used to say was **'and only through understanding can man heal'**. No amount of meditation, no amount of yoga, Reiki, whatever, all these different things, can heal us — he was very adamant about that. He said that's just creating another system. The moment you start necessarily saying 'meditation is going to give you all the answers' or yoga, yada, yada, yada, you're still not understanding the human condition. And having tried yoga and all these things, I'm like, 'yeah, that's very true inasmuch as I meditate and all these things, which calmed the mind like a relief mechanism, but it did not give us a healing explanation to the human condition'. And actually why, why do you even have to get over your mind? Why do you even have to stop your mind and overcome it and transcend it? Why can't we just understand it in the first place and through that understanding, and heal? And that understanding is definitely provided by Jeremy's book. It's incredible.

[151] **Tony**: It's just amazing to hear all that, Refentse. Honestly, it's just a such a clean take on it all. Where do you live? What's your background?

[152] **Refentse**: I live in the capital of North West [Province], Mahikeng. I'm from a Tswana background. I went to the school here, at the International School of South Africa, which is a pretty good school here in my town. A fortunate thing is that since it was an International School I grew up with many different people, of many different races. I played cricket growing up — I was pretty good at it, just short of playing for the under-19s South African team. When I look back at it now, that was one of the best experiences I've had, because I grew up with many different people — Afrikaners, Englishmen, Indians, other black people. It was a very nice melting pot. So I really got to experience many different sides of life, if I may put it that way.

[153] But as I got older, from probably Form Two [age 14], that's when my love for sports was still there, but that's when confronting the human condition became more and more of a thing. That's why I started reading more, being more isolated, spending more time alone. That's when the real change happened. That's when I started writing a lot and I'd say it was that period of time, from 13-14 onwards, when I really started dedicating my life to writing, to becoming a writer. And when I look back at it now, the main reason I became a writer was simply because I wanted to find out the truth. I wanted to understand the truth. I wanted to get to grips with it and reading, writing and questioning were the only three ways that I really knew how to do it. I didn't know any other way on how to go forward. Because every time I would ask myself questions, no one around me had the answers. So my friends didn't have the answers, my parents didn't have the answers, my teachers didn't have the answers. So I was like, 'Wait, where can I go for the answers?' And the one place I found was books. So I started reading the classics, you know. *The Great Gatsby*; *The Idiot*, *The Gambler*, all these Dostoevsky books; Bukowski; Hemingway; Albert Camus; Samuel Beckett. I really went down that rabbit hole. Aldous Huxley. All these guys, I even read [Darwin's] *Origin Of Species*, but it was just too thick for me and I couldn't really understand it at that time when I read it. I even went into astronomy, watching *Cosmos*, reading *Cosmos* by Carl Sagan; Neil deGrasse Tyson; Stephen Hawking, as well, his book *A Brief History Of Time*. I was really trying to reconcile everything. That was why in Reginald's [WTM Eastern Cape Centre] video when he was explaining his life, trying to connect this piece and that piece and that piece, I related because I really tried to do the same thing and it was just so difficult. It was just so difficult. That's why having seen Jeremy do it, it's absolutely mind blowing.

I'm like, 'wow!' For one man to be able to really go that whole journey and write it down in such a clear, concise way. Whoa! I cannot understate the point of what Jeremy did. There are no words that can describe that book, no! *FREEDOM* is the ultimate. There is no other book like that. That knowledge, I gave it hands down, even 10 out of 10 is not enough.

[154] Honestly we're living in the greatest time of man's history. The fact that we are living in the time with the explanation having been found, that's just beyond words to be honest. And just understanding Christianity—even that I got from *FREEDOM* and it's so incredible because, growing up, my grandmother was such a deep Christian. She used to make us go to church, all my cousins, and at that time I just didn't understand. It felt so forced. Even as Jeremy was saying in the book, our religions first started with monotheism and then prophet-centred religions from Abraham, Moses, and then Jesus going forward. Now it makes so much sense that to a large extent, once I got out of religion and became an atheist, I sort of disliked religion and looked down on it for a while; I was like, 'It's almost foolish, who would believe such a thing?' But once you really understand it, those words, everything Jesus and others in the Bible said, it's profound beyond measure and just even to think that Jesus actually made the biggest sacrifice one man could make for truth—I can't even wrap my mind around how huge of a sacrifice that was just to emphasise Integrative Meaning and all that. Wow. It's truly getting out of [Plato's] Cave. Getting out of that cave, it's amazing. [See F. Essays 38-41 on Christ explained and the demystification of religion.]

[155] So starting a Centre is something I've been giving a lot of thought to [and Refentse has since opened the Mahikeng WTM Centre in South Africa]. Going forward that is something I would

really like to do, because once you understand the human condition there is really nothing more; there is nothing more important than the human condition to be honest, so I'm willing to dedicate a good chunk of my time towards it.

- - - - - - - - - - - - - - - - -

[156] You can see many more relieved and inspired responses just like Refentse's in the VIDEO ENDORSEMENTS section on our home page at www.HumanCondition.com.

Video Endorsements

www.ingramcontent.com/pod-product-compliance
Lightning Source LLC
Chambersburg PA
CBHW051432270326
41934CB00018B/3480